Come hither, boy; come,
Come, and learn of us . . .
Thy grandsire loved thee well:
Many a time he danced thee
on his knee,
Sung thee asleep, his loving
breast thy pillow:
William Shakespeare
Titus Andronicus

I was their tutor to instruct
them . . .
As sure a card as ever
won the set . . .
They learn'd of me
William Shakespeare
Titus Andronicus

Have I not here
The best cards for the game
William Shakespeare
King John

PREFACE

Children love playing. They can play and play for hours on end. They don't seem to tire, and would even ignore calls for "bath-time" and "dinner", and go on playing.

Many children don't like work, especially school work or homework. In particular, many children don't like math. They find math difficult and incomprehensible. Reluctantly, they struggle at it, and endure the boredom, difficulties and unpleasantness. Sometimes when they do not fully understand what they are doing, they end up frustrated, and "hating math" in the process.

But math is vital in education, and students have to study it until the 10th grade (O-level). Math is also a "hierarchical" subject — it means that if you do not understand or know how to do math at lower levels (the foundations), you will have even greater difficulties at higher levels.

My five-year old granddaughter Rebecca loves math. She can do mental and written sums for hours on end. She would even ignore "bath-time" and "dinner" calls, and continue doing her math.
Except that she doesn't regard what she is doing as "math".
She thinks she's "playing cards" with Grampa!

XI

The games in this book are suitable for both boys and girls. For convenience, we will use the pronoun "he" in this book, instead of the more accurate "he or she".

I am a grandfather, with a special interest in teaching math to the young. I love my grandchildren; and I love math. So, I have read many books and articles on math; on the teaching of math; on "the math brain"; "the math gene"; "math puzzles"; "math circus"; "math delights" etc. I have spent many hours thinking about how to make math easy and fun for my grand-children.

Last year, I was at a conference overseas. One day, over lunch, I happened to sit next to a mother with young children. The conversation drifted to the difficulties of bringing up young children. She confided that her children in primary school "couldn't do math"; "couldn't add". I shared my experience with her, about how my then five-year old granddaughter could add and subtract triple digit numbers. She was surprised. I explained to her how we did it.

On another occasion, I was having a business lunch with some company directors, including two university professors. The subject of conversation drifted to the book that I was writing. One professor had a six-year old daughter. "She's very bright. But from time to time, she has difficulties with her math." The professor wanted to help her overcome her frustrations when she has her difficulties in math. I explained briefly how I "played cards" with my grandchildren. "You must hurry up and finish your book quickly. I'll buy a copy from you right away."

The thought that there may be other young children with similar difficulties with simple math prompted me to finish writing this little book, to help in a small way, children with similar difficulties. My hope is that parents and grandparents would read this book, spend some time "playing cards" with their young loved ones, and help them, in a fun and pleasurable way, to overcome their fear of and difficulty with math.

I am of the opinion that how you teach is more important than what you teach. Hence, whenever possible, parents and grandparents should play with their children. Teaching through play is a time-honour approach. Do not delegate this critical function to teachers, especially the teaching of this "difficult subject, math". With limited time and resources at their disposal, teachers have to deal with classes with dozens of children, with different abilities.

Children, even very young children, enjoy competition, which gives an added sense of achievement when they "beat" some one else. (My grand-children love "beating" Grampa in our card games. In their innocence, they know they are not "as clever as Grampa" but when the results showed that they had "beaten" Grampa, they are absolutely delighted, and tell everyone in the family about their achievements. Such achievements help them develop a healthy sense of self confidence and a positive self image in a world of gigantic grown-ups.

Children also learn through challenges — starting with the super-simple, and progressively, step-by-step, moving upwards, to higher and higher levels. A game (or task) that is too simple for a child soon becomes boring. Conversely, a game (or task) that is too difficult soon becomes frustrating. Therefore parents and grandparents must always bear this in mind. When a game becomes too easy or boring for a child, create your own variations on the same theme, or move upwards to a new game with a higher level of intellectual challenge. The art of keeping the child enjoying his games (and his learning) is to balance the ability of the child with the intellectual demands of the game at hand. Diagrammatically, this can be represented in the figure below.

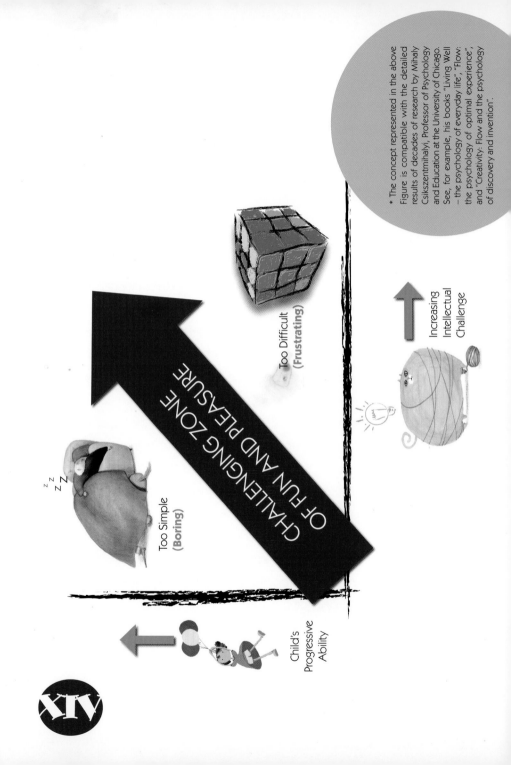

XIV

Too Simple
(**Boring**)

CHALLENGING ZONE
OF FUN AND PLEASURE

Too Difficult
(**Frustrating**)

Child's
Progressive
Ability

Increasing
Intellectual
Challenge

As the child's abilities progress, move slowly up the ladder of intellectual challenge, always listening carefully to the child's preferences, for clues on how he's enjoying the game, and learning the math. Never rush a child when he is not ready, or does not fully understand the game, or is unable to do the task required. Do not push him to play the game that he doesn't like.

Even as you encourage the child to play the games at progressively higher levels, be sensitive to the child, and let him decide what he wants to play. This way, "math" will always be "play" to him.

The value of an education . . .
Is not the learning of many facts,
But the training of the mind
Albert Einstein

It is the supreme art
of the teacher
To awaken joy in
creative expression
and knowledge
Albert Einstein

How you teach
Is more important
than what you teach
Y E O Adrian

The best teacher is not the
one who knows most,
but the one who is most
capable of reducing
knowledge
to that simple compound of
the obvious and wonderful.
H. L. Mencken

The most important job
in the world,
second only to being a
good parent,
is being a good teacher.
S. G. Ellis

In a completely rational world,
the best of us would be teachers . . .
because passing civilization along
from one generation to the next
ought to be the highest honour
and the highest responsibility
anyone could have
Lee Iacocca

INTRODUCTION

My grandchildren learn math by "playing cards" with me, beginning when they were 3 and 5 years old, respectively. In this book, I have listed some of the games we played.

The games are broadly grouped into three overlapping levels of intellectual challenge, based on my survey of children's math books, from nursery to Primary 6 (Sixth Grade).

I   Nursery and Pre-School

II   Pre-School and Primary

III   Primary and Higher Primary

Parents and grandparents must always remember that children develop at different pace. Some make steady continuous progress; others develop in spurts, with pauses in between. (Albert Einstein was well-known for being slow, in beginning to talk as a child, and later, slow again in school work!). So parents and grandparents should not be unduly worried. Nor should they make unnecessary comparisons with other children of similar age groups.

Parents/grandparents should feel free to pick and choose games from all three categories, and play with their children, regardless of age. If a child does not like a game or is uncomfortable playing with it, change the game to one that he likes. By being sensitive to the child, parents/grandparents can gauge their child's level of development. My four year old granddaughter Kathryn plays games in both Sections I and II. She loves "open-ended" sequencing (see Game 6), "fishing" (see Game 9), "mini-gin" and "SNAP". Rebecca, aged 6, plays games from all three Sections. "Fishing" (Game 9) with all its different variations, and "Logic" ("Tee Taa Too", Game 16) are among her favourites, with occasional games of "SNAP" and "Gin Rummy".

The guiding principle always is that the child must enjoy the game he plays. If he is bored, or conversely, if he finds it too difficult, change the game. Always be guided by the child's preference. The last thing adults should do is to force the child to "play", "for his own good". If you do, then the card-playing would be counter-productive, and the child will not enjoy his "math" anymore.

Also, remember not to rush the child. Even within a single game, there are different levels of intellectual challenge, some easier than others. Different children grasp the concepts behind the different variations at different pace. A child may initially not fully understand the basics behind a game; then one day — out of the blues — everything becomes crystal-clear. This is the classic example of the "eureka" discovery.

Play with your child and watch him learn, day by day. You'll be glad to be party to his adventure in discovery. As William Shakespeare said: "Go, play, boy, play"

Nothing you do
for children is
ever wasted
Garrison Keillor

Children are our
most valuable
natural resource
Herbert Hoover

From the very beginning of
his education,
the child should experience
the joy of discovery
Alfred North Whitehead

Our greatest
natural resource
is the minds of
our children
Walt Disney

The Universe,
Nature's great book, is
written in mathematics
Galileo Galilei

We cannot hope that many
children will learn math
unless we find a way to
share our enjoyment
and show them its beauty
as well as its utility.
Mary Beth Ruskai

NURSERY AND PRE-SCHOOL

Recognition of
**Patterns, Colours
Shapes and
Numbers**

Game 1
Recognition of
PATTERNS
pg **5** >>

Game 2
Recognition of
COLOURS
pg **11** >>

Game **1**

Recognition
of

PATTERNS

GAME 1

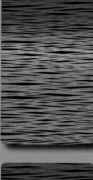
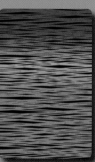

Using two different packs of cards, arrange the cards face-down in the manner shown in the Figure. Encourage the child to help you to do it; and then to follow your example; and finally to do it all by himself.

Once the child is able to do it consistently, use the same two packs, shuffle the cards to mix them up, and let the child arrange the patterns again from the random pack of cards. Congratulate him when he has done it.

Once the child has mastered this variation, increase the number of packs to three, and later to four, and have the child arrange them in three, and later, four blocks.

Again, after the child is able to handle four types of cards in a randomly distributed pack and arrange them in four square blocks, according to their patterns, create your own variations: e.g., have the child:

> arrange cards in 3 x 3 (9 cards each)
> arrange cards in circles
> arrange cards in diamond shapes, etc.

All the time, while the child is playing with his cards and patterns, make funny, loving, happy noises.

NEVER   scold a child who doesn't do what you want. ✕

NEVER   tell him that he is wrong; show him the right way. ✕

NEVER   discourage him. ✕

NEVER   force him to continue "playing" when he wants to do something else. ✕

NEVER   rush a child to do more than he can. ✕

ALWAYS encourage him. ✓

ALWAYS congratulate him when he has completed his game, or even part of his game. ✓

ALWAYS listen to him and his preferences. ✓

ALWAYS go along with the child if he is inventive and creative, and wants to play cards in a totally different way. ✓

NEVER

force him to continue "playing" when he wants to do something else.

Intellectual skills learnt from this game

The young child learns:

1. to recognise the difference between two, three and later on, four different patterns at the back of the cards.

2. to group them into different geometric shapes such as rectangles, circles, diamonds, etc.

3. to handle cards from a randomly shuffled pack and allocate the different patterned cards accordingly into the correct groupings.

4. simple psycho-motor skills and coordination.

Game **2**

Recognition
of

COLOURS

GAME 2

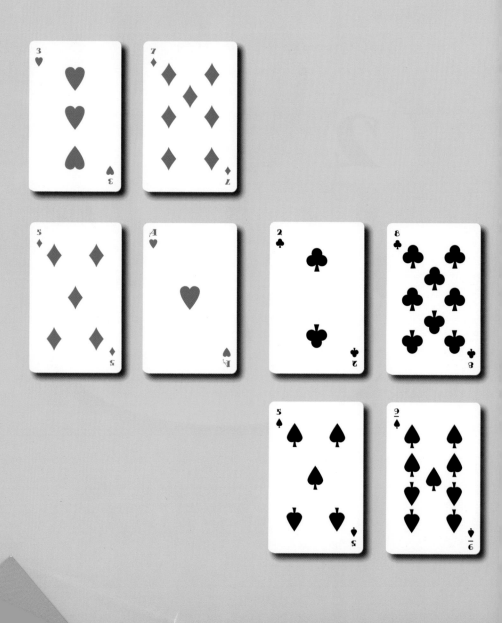

Game 2 appears superficially to be similar to Game 1, but there is a significant step upwards in terms of intellectual challenge. As in Game 1, arrange four red cards and four black cards in two rectangular blocks (see Figure).

The child learns to focus only on the colour and ignore the differences in integers , shapes and quantities of objects on the cards. (This is a valuable principle in math, to be able to extract the relevant characteristic of an item with multiple characteristics, and to ignore the rest.) In a typical card, there are at least four characteristics: colour, suit, integer and quantity of objects. Later a child will learn other more subtle characteristics, such as symbols (e.g. A for 1), sequencing (1, 2, 3, . . . 10), and flexibility (as in Joker for many different games).

The word "integer" will be used to refer to the number in the corners of the cards. The word "quantity" will be used to refer to the number of objects (e.g. hearts on the cards.)

Again, using the variations in Game 1 and others that parents/grandparents can create on their own, stretch the child intellectually while having fun. (For the first twelve games in this book, it is preferable that the "picture" cards be excluded, to avoid confusing the young child. Beginning with Game 13, we can introduce the "picture" cards; after that, when parents/grandparents should ever play any of the earlier games with the child, they can do it together with the "picture" cards.

NEVER

scold a child who doesn't
do what you want.

Intellectual skills learnt from this game

The child learns:

1. to focus on one characteristic of the card only, and
   ignore all the rest. Some children may wonder how
   you can group a card with one red heart with another
   card with ten red diamonds! Explain slowly. Once
   the child masters this ability, he would have
   developed and acquired a great intellectual skill that
   will stand him in good stead, not just for math in
   school, but also for life outside and beyond
   school.

Game 3

Recognition
of

SHAPES

GAME 3

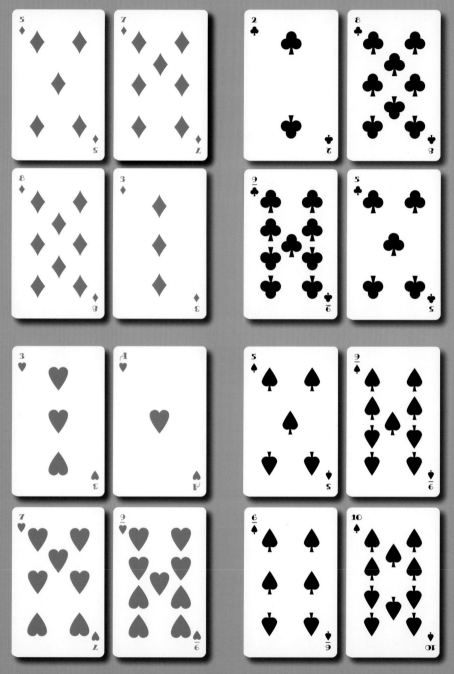

Once the child is able to group according to colours, move upwards and encourage the child to group according to suits. This requires him to separate the red hearts from the red diamonds, and the black spades from the black clubs (see Figure). In other words, his powers of differentiation are being sharpened, moving from colour to suits. Point out to him that the names of the four suits are spades, hearts, diamonds and clubs.

Always create your own variations to add interest to the game. Listen to the child; sometimes, remarks made by the child can give indications of areas that need further clarification or reinforcement. Adjust your game accordingly. Always keep the child's enjoyment paramount in whatever you do.

Intellectual skills learnt from this game

Over the past three games, the child learns to discriminate progressively minor differences — first, patterns at the back of the cards without distractions; then colours, and finally suits. To be able to distinguish minor differences when major characteristics, e.g. colour, are the same trains the child to be highly selective. The child learns to distinguish between two suits, both in red (or both in black); learns to ignore the difference in quantities e.g. red ace of hearts goes with red 10 of hearts, and not with red ace of diamond. While such distinctions may be obvious for an adult, for a child, this is quite an intellectual achievement.

ALWAYS

go along with the child if he is inventive and creative, and wants to play cards in a totally different way.

Recognition
of

NUMBERS

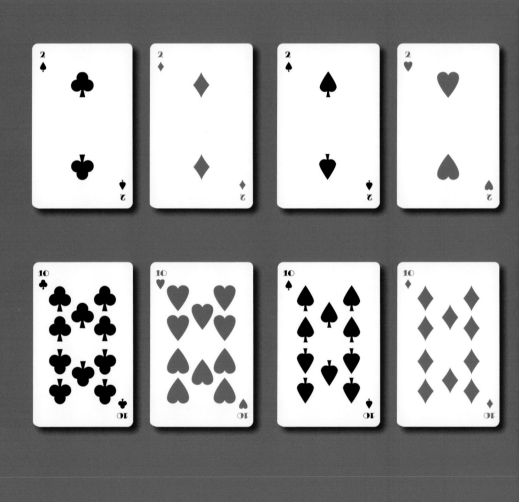

This game requires some preparation on the part of parents/grandparents. Take out all the 2's and the 10's (four cards each). Ask the child to arrange them according to the quantities of objects in the card. (see Figure). The child has to ignore all the characteristics except for the quantities, putting the 2's together (regardless of suit or colour) and the 10's together. The child is in the process of learning to associate three items — namely the sound "two"; the quantity signifying "two" and the integer "2" in the corner of the card. The same mental processes apply for "ten".

After the child is familiar with the 2's and 10's, progressively increase the intellectual challenge by moving to 4's and 5's; then to 6's and 7's and finally to 9's and 10's, where the differences in quantities are small. The child has learnt to be observant and concentrate on minor differences.

Soon, the child will be able to "read" the integers in the corners of the cards as he informally, "unconsciously" associate sight and sound of the numbers. [1]

We do not use aces in order not to confuse the child at this stage. Later the child will learn that A is used to represent 1, but this is only for cards and not in math generally.

Intellectual skills learnt from this game

For a young child, this is an extremely important game as he learns a number of complex skills such as:

1. counting the quantity of objects.

2. associating the quantity with the sound of the number

3. associating the quantity with both the sound and the integer in the corner.

4. doing all of the above while at the same time ignoring the more dominant characteristics such as colour and suit of the cards.

Do not rush this lesson for the younger children. Take your time, play this game often until the child is comfortable and does it well. (Remember it is not easy for a child to distinguish between 9 and 10 objects, or to group 9 red objects together with 9 black objects.)

Game **5**

Recognition
of

SEQUENCES

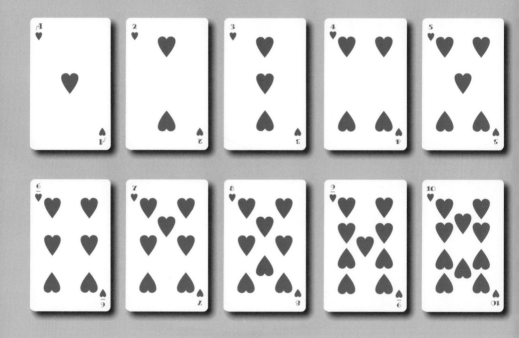

Extract all the cards of one particular suit (I use "hearts" whenever possible, because of its bright red colour as well as its very positive connotation of "love" for the child). Show the child how you can line the cards up, beginning with A (explain to the child that A means 1 in cards), 2, 3, . . . 9, 10. Then encourage the child to do the same (with cards from another pack), using your row as a model. It may take the younger child a while to understand. Be patient.

In the process, the child learns progressively, with positive reinforcement, to recognise both the quantity and the integer on each of the cards, and to arrange them in an orderly sequence. While this may be obvious to an adult, seeing a child work out in his mind, the intricate connections of a number of characteristics, is really an eye-opener for parents/grandparents.

In due course, all the complexities click in a child's head; and he starts doing the sequencing correctly. Congratulate him heartily. This is a rare moment for parents/grandparents; to see the intellectual development and transformation of the child right before your eyes. The feeling of pride, both in the child upon his achievement, and in the parents/

grandparents is truly priceless, and shows clearly why helping your child learn math the fun way is truly worth all the time and effort.

As you "play cards" with him over the weeks and months, and as he progresses upwards through the increasing intellectual challenges of the subsequent games, you will have many opportunities to witness his "eureka" moments. You'll be surprised, sometimes, at how fast the child learns, after initial attempts when he could not do it. Seeing a child put cards at random one week, and then, seeing the same child put all the cards in the correct sequence a week later, is a gift available to all parents/grandparents, who are willing to spend time "playing cards" with their loved ones.

When the child is able to sequence the "hearts", move on to another suit, say red diamonds, and then to the clubs and spades. When the child has little difficulty making the transition from one suit to another, it is clear that the child understands fully the significance of numbers, quantities and the stepwise increases of the sequences.

Intellectual skills learnt from this game

For a young child, this game is extremely educational and hence extremely valuable. So take your time, and do not rush the child in his initial attempts to put the cards in the correct sequence. If the child finds it frustrating, move to another game and come back to it later. Do not rush; do not force the child to "play" if he does not want to.

When the child is able to do it, sometimes almost effortlessly, he would have learnt:

1. the concept of quantities and integers and their association with numbers, with renewed reinforcement.

2. the step-by-step increase of numbers from 1 to 10. As he begins to count verbally, and to remember the correct order (or sequence) he learns also to associate with the quantity of hearts on the card. (Remember when a child first learns "verbal counting", he may just be repeating sounds that he has heard from adults — hence it is not uncommon to hear young children count "1, 2, 3, 7, 5, 10, . . ."

3. the difference of suits, and the similarity of sequencing, regardless of the change in suits — in other words, the sequence from 1 . . . 10 is independent of the suits and the objects on the card. This is the beginning of "abstraction", where the child learnts oneness, twoness, threeness. . . tenness as entities, without these entities being linked to objects such as "one dog" "three cats" or "ten fingers". This is a vital intellectual ability and is fundamental to mathematics, which deals with numbers in the abstract and not with cats and dogs.

4. a useful "mental reference framework" from 1 to 10. When the child can accurately and without effort verbalise (or sing-song) "1 to 10", he is now able to mentally compare the relative size and position of the single digit numbers. This will stand him in good stead when he learns later to count to 100!

Game **6**

Recognition

of

MULTIPLE
CHARACTERISTICS

GAME 6

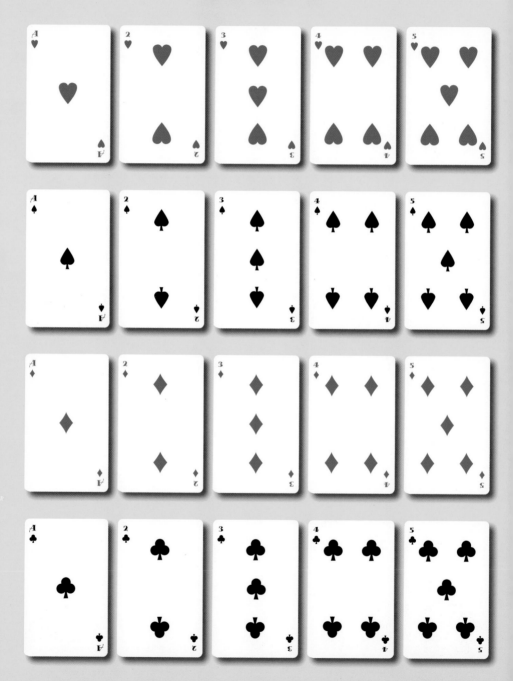

When the child is comfortable with the first five games and the different variations that parents/grandparents create for them, the child is ready for game 6 — which features the culmination of the multiple characteristics in cards.

Begin with one pack of cards. First encourage the child to sort the cards, faced-up, into four different groups according to their suits. Then encourage the child to sequence the "hearts", from 1 to 10. Then do the same for the other suits in turn. Congratulate the child as he is doing it.

When he is comfortable sequencing suit by suit, and doing it effortlessly, move up the intellectual ladder a little, by having all forty cards, faced-up, but spread out on the floor at random. Then help the child to sequence the four suits, beginning with "hearts" A, 2, . . . and so on. Congratulate him, and encourage him now to do it all by himself.

Again, when the child is comfortable, help the child to think of the position for the different cards, with the sequences mentally in his head. When he picks up a "10 of hearts", help him to put it in the top

GAME 6

Recognition of

MULTIPLE CHARACTERISTICS

34

right hand corner. When he picks up an "Ace of hearts", help him put it in the top left hand corner. If he picks up a "5 of hearts" help him put it in the middle at the "top row". As the child plays through the forty cards, help him to "fill-in-the-blanks". The child learns from your example and help, and in next to no time, he can be encouraged to do it all by himself from the forty random faced-up cards on the floor. (Parents/grandparents can also progress at a slower pace, if necessary, by using only 20 cards from two packs, either from two red suits, or from one red suit and one black suit.) Keep the game fun and lively (its beginning to be more and more intellectually challenging for the young child, so make sure that he's having a lot of fun doing it).

As the child progresses and gets more and more competent in filling in the cards in their correct suits and sequences from the faced-up cards, move upwards to the next step. Shuffle all forty cards and keep them in a pack, faced-down. Then "open" one card at a time, and ask the child to place it. When the child has reached the stage where he has a "mental reference framework" for the four sequenced suits, he can put that first card on the floor in a spot that he has designated. Then "open" up card two, and card three and so on. It is a real pleasure to see a child putting the cards in their approximate positions on the floor, and then adjusting them as new cards are opened up randomly. When the child is able to do this, congratulate him heartily. And congratulate yourself too, because the child has reached an exceeding important and valuable intellectual milestone in his life. The ability to conceptualise information and data, and to organise their positions relative to one another in his head, with a "mental reference framework", is a great achievement. Hence, make sure that you are patient, and do not rush the child in playing this game.

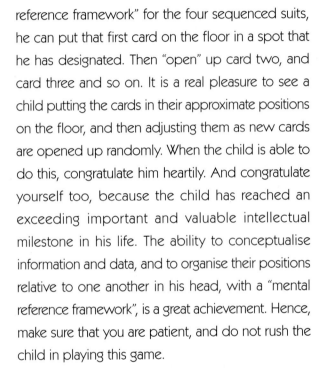

Intellectual skills learnt from this game

With the completion of this game, the child has learnt a coherent array of intellectual skills which demonstrates tremendous achievements — mentally creating a "four-by-ten matrix" in his head, understanding and knowing the correct sequence of the numbers from 1 to 10, understanding the separation of suits despite their similarities of colour and shapes of objects, and knowing the step-by-step orderliness of numbers and quantities.

Much of an adult's intellectual activity in normal or working life requires this ability. New information that we receive in large quantities day by day has to be slotted into our "mental reference framework" before we are able to assimilate the new information, and learn from it. A person who can do this effortlessly "learns" much faster than one who cannot associate the new input with his existing knowledge or data base, hence making retention, recall and remembering so much more difficult. (A good illustration of this phenomenon is the established observation that good chess players can better remember the positions of chess pieces in the middle of a game (after they have been removed from the board), than non-chess players. This is because the chess players register the positions of the pieces in a mental framework that they have in their heads. Such mental frameworks for chess pieces are absent in the heads of non-chess players!)

With the completion of Game 6, the pre-school child is ready to embark on more arithmetical and numerical games. As a point of interest, even though very young children learn about patterns, and older children learn more about numbers in their math lessons in school, mathematics, especially advanced mathematics is often regarded less as "a science of numbers" which is too restrictive, but more as "a science of patterns". This is because many branches of advanced mathematics deal more with patterns. The numbers are only a tool for manipulating the equations for the patterns. Branches of mathematics which deal largely with patterns include fractal, topology, knots, and networks. Even a strongly arithmetical branch of mathematics — number theory — deals with patterns in numbers. One of my favourite subjects — infinite series — is all about numerical patterns.

Whenever you can,
Count
Sir Francis Galton

People who can't count
Don't count
Anatole France

PRE-SCHOOL AND PRIMARY

Simple
Arithmetical
Concepts

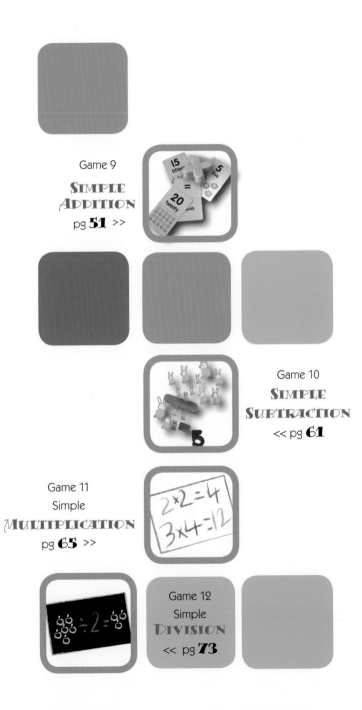

Game 7

"GREATER THAN"

GAME 7

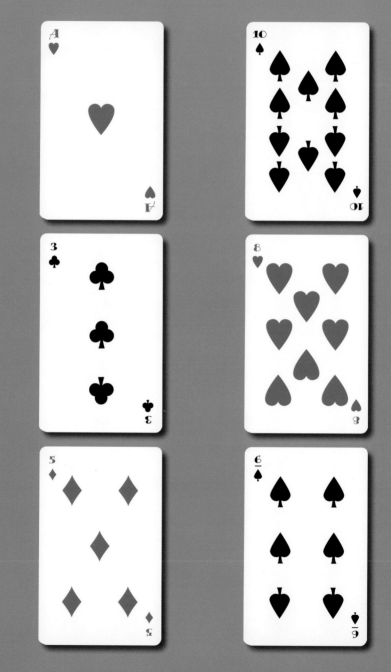

"GREATER THAN"

Games 5 and 6 informally and intuitively introduced the child to the concept of quantities, with some cards having more objects than others. In Game 7, we want to make explicit the concept of "greater than", a important mathematical concept used extensively in later years in school.

Start by showing the child two cards; say an "Ace of hearts" and a "10 of spades". Ask him, "which is greater?" If he gets it right, slowly reduce the difference between the two cards until you offer him 4 and 5; 6 and 7; 7 and 8; 9 and 10. In these pairs of cards, the quantities of objects in the cards are very similar, and are arranged in similar patterns. If the child has little difficulty, and answers the "greater than" questions quickly, you can infer that the child can read the integers and know the concept of "greater than" well. If you wish, you can "test" him by showing him only the integers in the corner of the cards. If he can read the integers consistently, then you know he has mastered his basic integers which is extremely important when he goes to school and later deal with double and triple digit numbers.

"GREATER THAN"

Create your own variations e.g. use a "9 of hearts" and a "10 of diamonds" and ask "which is greater"; use a "7 of spades" and a "8 of clubs"; use a "6 of hearts" and a "7 of spade" and so on. The use of different colours, and suits serves to add intellectual challenge to the game. Over time, the child learns to ignore these "non-relevant" characteristics and focus only on the critical ones.

ALWAYS

congratulate him when he has completed his game, or even part of his game.

Game

8

"LESS

THAN"

Game 8 is the converse of Game 7. Most children who can play Game 7 will have no difficulty with Game 8. Again it is useful to play Game 8, to ensure explicitly that the child understands the concept of "less than"

Begin the game slowly as in Game 7, and add variations as the child becomes comfortable with the concept.

This may also be the time to introduce the "time element" which children love in games. The game "SNAP", an extremely simple game, can be used together with many of the games in this book to add interest and excitement to the learning of math the fun way.

Begin simply with the very young child by "opening" two cards simultaneously (one by the child and one by the adult). Encourage the child to say "SNAP" when he sees two cards of the same colour. Most children enjoy this super-simple game immensely. After

 We use the terms "greater than" and "less than" instead of two similar terms "more than" and "smaller than" because the former two terms are the standard terms used later in math in school and in university. This ensures that the child does not have to unlearn unconventional terms in future.

a few rounds to ensure that the child understands the concept, progress upwards to "two cards of the same suit". Here a bit of preparation may be necessary — take out the half set of cards of the same colour. Again, after the child is comfortable, play with the whole pack with all four suits. You'll be surprised how fast a child learns to "SNAP".

When the child is comfortable with the above variations, introduce "SNAP" for "greater than" — here after a child has said "SNAP", he must also say whose card is "greater than" the other. Play also the converse game "less than" to complete his math learning.

If parents/grandparents consider it helpful, they can introduce the element of "winning" by letting the child keep the cards he won by "SNAPping" at the right occasion. To discourage indiscriminate "SNAP's", introduce the idea of a penalty — a wrong "SNAP" will make a forfeit of the two cards to the opponent.

GAME 8
"LESS THAN"

This new challenge with an element of time and competition will inject great interest into the games, and the child will enjoy "playing cards" even more.

When the child is ready, parents/grandparents can encourage the child to count his winning cards (if necessary, with a bit of help from the adult). Then encourage him to write down his "score". Slowly, the child associates the numbers that he writes down with the cards that he had won. Parents/grandparents can set the example for the child to follow.

All these variations, together with others that parents/grandparents can create on their own, will ensure endless fun for the family playing together, and learning math at the same time.

15
fifteen

5
five

=

20
twenty

quals

Game 9

Simple

ADDITION

Simple ADDITION

This is one of the most important games in this book, and can be the basis of a lot of math learning. Addition creates the foundation for math excellence, and hence special attention to a child's mastery of addition is essential for future progress in math.

The basic game here is commonly called "fishing". But we will use it with many different variations to help the child master both mental as well as written addition.

As always, begin slowly with the basic game. Using forty cards from one pack, give ten cards each to the child and to the adult, with another ten cards faced-up on the floor (I prefer playing with my grand-children on the floor because the table is often associated with work, whereas the floor is often associated with play). Then, taking a card from one's hand, we "fish" for a card on the floor, such that the total of the two cards adds up to 10. If a player cannot get two cards to add up to 10, he draws another card from the reserve stack (of ten cards), and tries to fish again for 10. If he is still unable, he leaves the card on

Children as young as four can play "fishing", fishing for a total of 2, 3, 4 or 5, with the total increasing slowly as they become more conversant with addition.

the floor, and the turn goes to the other player. Keep the cards that you win by "fishing".

When all the cards on the floor have been "fished" out, or no more "10's" could be made, the game stops, and all remaining cards in hand are returned to the central pool. Encourage the child to count his "winning cards". Then, if the child is able, encourage him to write down his "score". (Parents/grandparents can do the same to encourage the child, and set an example for him.) If it is available, use paper with medium-sized squares (see Figure on the next page), so that the child learns informally to write numbers in proper rows and columns. After two games, with two numbers recorded, one on top of the other, encourage the child to "add up" his winning cards. (Parents/grandparents should "arrange" it such that the child "wins" most of the time (i.e. has the higher score)! You'll be surprised how happy the child is when he "beats" his parents/grandparents at the game. This sense of achievement is immense, and is of extreme importance in helping a young child grow in self-confidence in a world of large grown-ups.

Addition is the foundation of Arithmetic, and Arithmetic is the foundation of Mathematics.

YEO Adrian

The writing of the numbers and the addition were all done by Rebecca, aged 5. (The names "BECKY" and "GRANDPA" were written by Grampa)

When the child enjoys playing the game, and writing down the score, and adding them up game after game (if necessary with some help from the adult), encourage him to be "the official score-keeper", and keep the scores and add up for both players. Being the official score-keeper enhances his sense of importance again, and reinforces his self confidence further. Learning how to add, while playing cards with parents/grandparents is truly easy, fun and pleasurable. You'll be surprised when the child wants to do this kind of math for hours on end, all the time "doing mental addition" when he's fishing, and "doing written addition" when he is keeping score. And being delighted all the time when his score is higher than the adult's!

Remember, addition is the foundation of arithmetic, and arithmetic is the foundation of mathematics. All the practice that the child gets playing this game will stand him in good stead, for the rest of his school years and in his life beyond school.

Using paper with medium-sized squares (preferably in an exercise book), encourages the child to write down the numbers in their proper boxes. With winning scores in double digits (tens and twenties), the child gets to learn informally what schools will later teach them as H.T.U's (namely hundreds, tens, and units). Simple addition of digits column by column gets the child used to addition of double digit numbers. After a few games, his addition gets him to triple digit numbers. The child learns effortlessly and informally, double digit addition and triple digit addition, and discovers that they are as simple as single digit addition! No more fear of "large numbers. (Figure — shows examples of Rebecca's score-keeping, when she was 5-years old).

Simple ADDITION

When the child is comfortable fishing for 10's and keeping score, change the game a little, and fish for "11"; then for "12", "13" and so on up to "20". Keep watch over the child's addition as you progress upwards. If he does not find it easy to add mentally to double digits, go slow, and stay at "11's" and "12's" until he is comfortable. Do not upset the child by moving to the higher teens too rapidly.

If after a few days of "fishing" and the child is very comfortable, progress upwards and change the game a little again — e.g. a total of "17" can be attained by the use of multiple cards (e.g. using either 2, 3, 4 or even 5 cards). Encourage the child to add aloud e.g. say "2 plus 5 equals 7; 7 plus 10 equals 17". Soon, you'll be surprised at the speed of his mental arithmetic as he zips along in his addition.

Again, when the child is comfortable, move upwards slowly to the twenties e.g. "21", "22", "23" etc. Remember, always keep an eye on the child, and ensure that he's enjoying himself. This is most important. Don't force him to "fish" for high numbers if he is not comfortable.

Once a child is able to "do mental addition" in fishing, and "do written addition" in score-keeping, he's going to find math in school "so simple" as my granddaughter always tells me. (Remember, your child may still not always get 100 marks, as children can sometimes get distracted, or are "careless" or misread or misunderstand questions. Getting a child to get 100 marks all the time is not as important as helping the child to develop a love for math.)

Parents/grandparents can also play "SNAP" with addition as a basis — if two cards add up to "10" (or "11", or any pre-arranged total). Injecting the time element into addition creates a lot of fun and pleasure, as occasional "mistakes" of "SNAP" are shouted out. (If considered useful, a penalty of losing the two cards can be imposed for "wrong SNAPs" to sharpen the game.)

Other variations that parents/grandparents can create include "open SNAP" i.e. saying "SNAP" followed by the total of the two cards e.g. "SNAP – 19" for a "10" and a "9". Parents/grandparents should remember to play along with the child, to allow him sufficient reaction time, and to win most of the

Simple ADDITION

"SNAP's". Keeping the game reasonably competitive will sharpen his mental prowess in addition, and will improve his score.

This game with all its variations gave Rebecca the greatest of pleasure, as she does her "mental math" and "written math" for hours on end, playing cards with Grampa. She did a lot of math before she went to Pr. 1, without thinking of it as math at all!

Fishing games are endless, and parents/grandparents can create their own (e.g. one physical game that my two grandchildren love is to slide down my outstretched legs. To "pay" for the ride (as they know from their experience at the games arcade), they must get me cards from a pool of "faced-up" cards. I'll say the number e.g. "19" (I usually use the high prime numbers as I noted that the other numbers are easier for Rebecca), and she will find the cards that add up to "19", say the sum aloud, and put the cards into my T-shirt pocket, pretending that it's a slot for the money, and then slide down my legs. Kathryn 4, gets the easier single digit numbers before she does the same.

Have fun with your own creative inventions.

Intellectual skills learnt from this game

Together with its different variations, the game of "fishing" (addition) is one of the most important games in this book, if not THE most important. With the myriad of possibilities, the child can play this game endlessly without feeling bored – in other words, he can do his math without feeling bored. On the contrary, he'll have lots of fun doing math, which is what mathematicians feel when they "do their math". Professional mathematicians are among the luckiest people in the world – they get paid for having fun and enjoying themselves, indulging in a fantastic intellectual game called "mathematics".

The child will learn effortlessly, informally, and with lots of love, fun, and pleasure, playing together with parents/grandparents and learning how to add, initially, single digits, both mentally and in proper written form, accurately and speedily. Over time, they can add mentally double digit numbers, and through score-keeping triple digit numbers. In the process, they overcome what for many children is the phobia for "large" numbers. Knowing how to add

Simple ADDITION

systematically one column at a time, and "carrying" forward the tens to the next column, gives the child mastery of the most basic of arithmetical skills. The child also learns good math habits e.g. checking his addition after he has completed it, to make sure that "carelessness" had not slipped in. When the answer is the same the second time, more likely then not, the addition is correct.

Once the child is able and confident in his mental addition and later with his written addition, he will be better off than many adults who "freeze" when it comes to addition. With a strong foundation in adding, acquired "effortlessly the fun and pleasure way", he will be in good stead for future work in arithmetic, and much later in the other aspects of math.

Truly as the Chinese proverb says:

"Teach a child to fish, and you feed him for a lifetime".

Simple

SUBTRACTION

5

Simple SUBTRACTION

For adults, substraction is just the converse of addition. But for a child, the two concepts, though related, are different; and it may require some mental adjustment for the child to understand the concept of subtraction fully.

Begin slowly again by playing "fishing", this time with a difference of, say, 1 — e.g. using a 7-card to fish for a 6-card on the floor. Encourage the child to say "7 minus 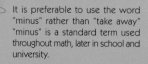 6 equals 1". Then fish for "2", and then "3" and so on. If you wish, allow for "reverse fishing" where a "7" on the floor can be fished by a "6" in hand (as long as the child say "7" minus 6 equals 1"). Children love variations, and such variations teach them flexibility, so that the child does not follow rules rigidly, especially when such rigidity is unnecessary.

Variations with "SNAP" can also be created and played.

When the child is good with his score-keeping (from Game 9), encourage him to do written subtraction by writing down his score and subtracting that of the

It is preferable to use the word "minus" rather than "take away". "minus" is a standard term used throughout math, later in school and university.

adult's. Do the subtraction on a separate piece of medium-sized square paper. The result will show him how much he has "beaten" his parents/grandparents by in the game. Help him initially, if necessary. He will enjoy this part of score-keeping immensely, especially on the occasions when he beats the adult's score by a big margin. Help him to subtract a "big digit" from a "small digit" (e.g. 6 − 9) by "borrowing one from the next column on the left" i.e. adding 10 more units to 6 to give 16 − 9. (Remember to cancel the digit on the left column and replace it by a digit that is 1 less than the original). When the child is motivated to find out by how much he has "beaten" his parents/ grandparents, he'll learn subtraction very quickly! And he'll enjoy it too. (You'll be surprised — see for yourself).

As with addition, create your own variations to add interest and excitement, and enjoy seeing your child master his math and wanting to do more. You can get a child to do written addition and written subtraction for hours on end, and with the child enjoying it.

Intellectual skills learnt from this game

Addition and subtraction are extremely important basic skills in math. And they are highly relevant in school and in life. Going shopping requires simple addition and subtraction all the time. People who cannot add or subtract have no choice but to accept what the shopkeeper says when buying multiple items, and to accept whatever change given back to them, assuming that the shopkeeper is honest and correct in his arithmetic.

After a child is proficient in his addition and subtraction from his games in cards, play shopping games with him. Then bring him shopping and encourage him to add for you, initially, keeping it simple by rounding off the prices of items to dollars. Then ask him how much change is expected for the $10 or $20 that you are going to pay the bill with. (The sense of achievement that a child feels when he can "participate" in real life transactions will be immense and will help him in his social and emotional development). His "playing" with cards can stand him in good stead and help him in his school-tuckshop and supermarket transactions.

Game
11

Simple

MULTIPLICATION

Multiplication is simply repeated addition, e.g. "3 times 2" is the same as "2 plus 2 plus 2". Yet most children have difficult with multiplication. Many schools resort to rote-learning by having the children chant the "times-tables".

Playing cards can be used to help the child learn multiplication the fun way. By a special variation of "fishing" where only identical numbered cards can be fished out, we can get the child to learn his "times-tables" the fun way, with minimum stress and pressure e.g. the child must fish for the same numbered card as his own, and say aloud the sum: "2 times 2 equals 4", "2 times 3 equals 6"; "2 times 8 equals 16". Over a game or two, the child learns his "2 times-table" very rapidly as he is motivated to win at his game. Effortlessly, the "two times-table" becomes part of the child's mental framework. (Parents/grandparents can initially select the appropriate "small" cards from all four packs to increase opportunity for "multiplication fishing" e.g. sixteen 2's, sixteen 3's and sixteen 4's to start the game. Thereafter, when he is comfortable with "$2 \times 2 = 4$"; "$2 \times 3 = 6$" and "$2 \times 4 = 8$", add another sixteen 5's.

Do not rush into the higher levels of "multiplication fishing", as multiplication takes some effort on the part of the child (even though it may be easy for adults).

When the child is comfortable with the "small" cards, move upwards slowly by introducing the "higher" cards and taking out the 2's and 3's. Watch the child's progress with his multiplication. If he makes good progress without difficulty, you can introduce all the "higher" cards. If the child prefers to play with the "smaller" cards, stay at the level that the child enjoys.

When the child is comfortable, you can play with all forty cards in the single pack or with forty cards from two packs selected with the cards that you want more practice for the child. In all circumstances, be sensitive to the child's needs and preferences.

When the child is comfortable with fishing from "2 × 2" all the way to "2 × 10", progress upwards by introducing "3-times fishing" i.e. one card in hand fishing for 2 identical cards (Again, some prior selection of "small" cards from multiple packs will be helpful to keep the game interesting.)

As with the "2 times-fishing", progress upwards at a pace that the child is comfortable with, until the child is able to do "3 × 2" all the way to "3 × 10".

Once the child is familiar with his "two-times table" and his "three times-table" and can give the answers without any effort, the child is ready for an important concept in arithmetic "the commutative law", namely the equivalence of "two times three" and "three times two", i.e. the order of the digits for multiplication is not important.

As the child is fishing for "2 × 4" mention to the child that we can also say "4 × 2 = 8". The child can see for himself that it is true and so effortlessly assimilate this fact. As the child plays with "2 × 4", "2 × 8", "3 × 7", "3 × 9" etc, he is informally learning the lower levels of the higher times tables e.g. "6 × 2", "6 × 3", "7 × 2", "7 × 3" etc.

Once a child is familiar with both his "two-times table" and his "three-times table" as well as his lower multiples of the bigger digits, he can now be introduced to the "easy" times table for 10:1 × 10 = 10, 2 × 10 = 20, 3 × 10 = 30 etc. You'll be surprised how fast he learns his "ten-times table"! Progress upwards

with the "five-times table". This may require a bit more time than for the "ten-times table".

Once the child is familiar with the game of "multiplication fishing" and his "five times" and "ten times tables", you can then gradually introduce the remainder of the "times-table" four, six, seven, eight and nine. Do not rush. The child has already assimilated the concept of multiplication; just add to his mental reference framework.

Allow yourself and the child some weeks of "multiplication-fishing" to learn his times tables. (For young children I prefer to stop at ten-times table. There is no need to go to the traditional "twelve-times table". They were necessary in the days of "imperial" measures, where "12 pennies make a shilling" and "12 eggs to a dozen". In the modern computer and calculator age of metric measurements, the ability to multiply mentally up to ten times is good enough for the purpose of mental development and discipline).

When the child is comfortable with his multiplication, and enjoys "multiplication fishing", progress upwards with another variation. With only

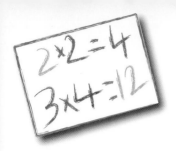

forty cards from one pack, play "open-ended multiplication fishing" where a card in hand can be used to fish any other card, as long as the child is able to say what the product is e.g. "2 × 9 = 18", when using the 2 to fish for the 9; "7 × 3 = 21", when using the 7 to fish for the 3; etc. You'll be amazed when you see your child do his multiplication referring only to his mental reference framework of his "10 × 10" matrix! (And all these achievements without the drudgery of sing-song times-tables!)

"SNAP" can also be used to help with multiplication fun. Throw out the two cards as before; say "SNAP – 3 × 8 = 24" if the two cards are 3 and 8; and so on. A lot of fun — a lot of learning — a lot of laughter — and make sure that the child has a lot of winning cards! He'll love his multiplication! You see for yourself how good he is!

Create your own variations to add more fun and laughter. The child will be pleased with himself and his achievements. And you'll be pleased with yourself, and with your child and his ability and performance.

Intellectual skills learnt from this game

Even though calculators nowadays are cheap and easily available, the ability of do simple multiplication up to 10"s is a valuable mental skill that is very helpful in a child's development. Through this game, the child is learning a major math skill without the need for boring "mindless memory work" through endless drilling of the times-table. He also learns "the commutative law" informally, knowing that "2×4" = "4×2" = 8.

ALWAYS

listen to him and his preferences..

Game
12

Simple

DIVISION

To adults, division is just the converse or opposite of multiplication — "no big deal". But to most children, division is one of the most difficult of the arithmetical skills that they have to acquire. It is even more difficult then multiplication. Hence parents/grandparents should take their time and introduce the "division" game to their child gently, slowly, without rushing him into it until he is comfortable with his addition, subtraction and multiplication. "More haste, less speed" will apply if the child's foundation with the three earlier skills are not strongly grounded.

Take the "division" game slowly, one step at a time. Always be guided by the child. If he doesn't want to "play" the division game, don't force him.

Begin by playing the basic multiplication game fishing for identical numbered cards. (e.g. a 7 fishing for a 7, saying "2 × 7 = ?". When the child answers 14, congratulate him on his correct answer. Then taking the two 7's in two hands (with one card in each hand) ask him "14 divided be 2 equals?" The child may or may not understand. Go through the "dividing" action again — put the two 7's together, say "14" and then separate the two cards into the two hands, saying

"divided by two equals ?" If the child answers 7, congratulate him. If he doesn't, go through the process one more time. He is very likely to get it this time. Congratulate him. Now try it with 8's. (If the child is still unable, you may wish to go back to "two 2's" for "4 divided by two equals ?" Be sensitive to the child. If he is not too keen, move on to other games, and come back to division some days later. The child may just not be ready. There is no hurry. Schools won't be doing "division" until after a few months of simpler math learning.)

As long as the child is happy with the game, play with division by two until the children is familiar with the concept of division.

Only when the child is comfortable should parents/grandparents move on to "division by three" games. Here it is useful, if two adults can play together with the child so that the child can see the physical division of the cards to three people. Remember, what is obvious to adults is not necessarily obvious to a child, so do be patient, and do not rush the child. Remember also, the primary objective is to have fun with the child. The math will follow naturally in its own good time. So don't be anxious when the child doesn't get his answers.

It cannot be over emphasised that division is a difficult arithmetical skill, so do not rush to "division by four" until the child is ready. Intersperse his division game by playing with multiplication from time to time, to refresh and reinforce his multiplication reference framework. You may find that it is helpful to the child to play "multiplication by three" followed by "division by three".

ALWAYS

encourage him.

NEVER

rush a child to do more
than he can.

Intellectual skills learnt from this game

The ability of a child to do division effortlessly is an invaluable skill in arithmetic. With the four basic skills of addition, subtraction, multiplication and division, the child has a very strong foundation for other aspects of arithmetic that he will encounter later in school, such as fractions, decimals etc. With these skills he will find subsequent requirements in other branches of math less difficult and more manageable, and will have less discomfort in math lessons in school.

I hear and I forget
I see and I remember
I do and I understand
Chinese proverb

A mind is a fire
to be kindled
not a vessel to
be filled.
Plutarch

Mathematics is a more
powerful instrument of
knowledge
than any other that has
been bequeath to us
Rene Descartes

PRIMARY AND HIGHER PRIMARY

Simple Mathematical Concepts

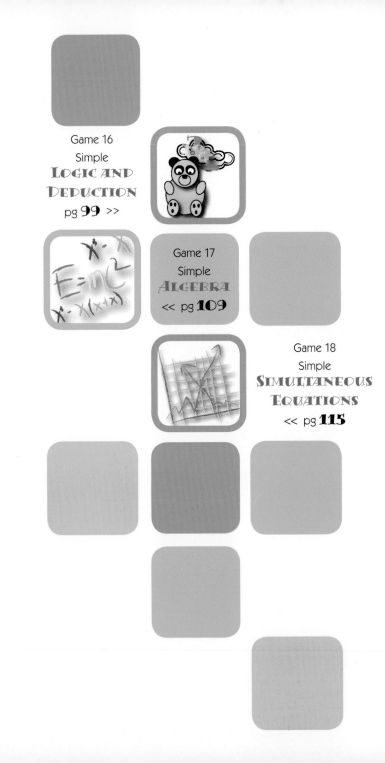

Simple

SYMBOLS

Up till now, we have been playing games with only the numbered cards, excluding the "picture" cards. In this game, parents/grandparents will use the picture cards and introduce the concept of symbols slowly to their child, in two stages:

1. Explain to the child that the "K" in the corner of the card is an abbreviation of "King". Point out the crown on the King's head, his luxurious clothes, etc; "Q" is for Queen; and "J" is for Jack; you can give a simple explanation that Jack is an assistant to the King and Queen.

2. When the child is familiar with K, Q and J, explain that for some games, J can be taken to represent 11; Q, 12; and K, 13. You can now reintroduce A which stands for 1, so that the child is familiar with all the symbols in the suit.

Then get the child to play "Game 5, Sequencing", this time including the J, Q and K. The child should have little difficulties in going from A, 2, 3, . . . 9, 10, J, Q, K.

After the child is able to do his sequencing up to J, Q, K without difficulty, you can go back to basic "fishing", but now explaining that for this game, J, Q and K are to be treated as 10's for purposes of adding!

Some children will ask about the difference between the two games "sequencing" and "fishing" and why J, Q, K are 11, 12, 13 for one and 10, 10, 10 for the other. Explain slowly, that these are different rules for different games and that the picture cards are used as symbols for numbers. Do not rush this concept of "symbols" and ensure that the child understands.

This symbolism of a "picture" card being used to represent different numbers in different games can teach an extremely important concept in mathematics. In an important branch of math, called algebra, the letter "x" is used to represent many different values in different equations and circumstances. Similarly the letters "y" and "z" can also be used to represent other variables.

The picture cards can be a very useful tool to reach this important concept in a fun way. Once, the child internalises this concept and uses it routinely in his games (they do, and in next to no time, too!) he will have less difficulty in his algebra later in school.

Intellectual skills learnt from this game

Symbolism is used extensively in math from school math to advanced math. Indeed mathematicians overwhelmingly use symbols, romanised alphabets, greek alphabets, graphs, diagrams, special signs such as brackets (e.g. for matrices) and special symbols like $\int$ (for integrals) etc. More complex combinations of multiple signs and letter are also used e.g. e, Σ, $\ln$, $\log$, d/dx, etc. With his mastery of games incorporating picture cards as symbols for different numbers in different games, the child learns informally a vital mathematical concept which will help him immensely in later years.

ALWAYS

go along with the child if he is inventive and creative, and wants to play cards in a totally different way.

Game
14

Simple

SETS

GAME 14

In mathematics, sets are a way of putting together different objects into different groups, classes or categories with common characteristics. The mathematical term for such groupings is "sets"

Cards are ideal for teaching children about sets in a play way.

In the figure opposite, we can group the eight cards into many different sets.

1. a set of three red cards and a set of five black cards.

2. a set of one "hearts"
    a set of two "spades"
    a set of two "diamonds", and
    a set of three "clubs".

3. sets of tens (e.g. 2 and 8; 3 and 7, 4 and 6, etc)

4. sets of non-ten (e.g. 3 and 4; 7 and 8; etc.

With a full pack, parents/grandparents can create their own games on the theme of "sets" e.g. sets of 10's using red cards only; sets of 10's using one red card and one black card only; etc.

The variations are endless, and they add interest and fun to the game, while helping the child understand and assimilate the concept of "set". Soon the child will understand that a single card can have a large number of different characteristics, enabling it to be grouped in different ways with other cards with common characteristics.

ALWAYS

congratulate him when he has completed his game, or even part of his game.

90

Intellectual skills learnt from this game

Sets are embedded in much of mathematics, and for the child, learning to make distinction between the numerous minor characteristics of a card helps the child to understand mathematics later when he encounters them in the form of:

$$x_1, x_2, x_3, x_4, \ldots$$
$$y_1, y_2, y_3, y_4, \ldots$$
$$z_1, z_2, z_3, z_4, \ldots$$

He will understand that grouping them for x_1, y_1, z_1 ... and grouping them for $x_1, x_2, x_3, \ldots$ are equally valid ways of mathematical manipulations for problem solving.

The child will also progressively appreciate the complexities of human nature — how a person can be good at one activity yet not necessarily be good in another. Many human problems arise as a result of wrongful typing of people based only on one or two characteristics.

Game
15

Simple

PROBABILITY

Simple **PROBABILITY**

Probability is one of the more difficult branches of mathematics, as well as in real life application of mathematics. Hence it is useful for the child to be introduced to the concept of "probability" the fun way through cards.

Start simply. Start with three cards, A, 2 and 3. Ask the child how many different ways he can arrange the cards to give a total of 3:

A + 2 = 3
            3

Congratulate him when he gets it correct consistently. Now try it with four cards. How many different ways of getting a total of 4 with A, 2, 3 and 4?

A + 3 = 4
            4

How many different ways of summing to 5 with A, 2, 3 and 4?

See for example, four simple illustrations and challenges in "Post-Appendix".

$$A + 4 = 5$$
$$2 + 3 = 5$$

Add another ace. How many different ways of summing to 5 with A, A, 2, 3, and 4?

$$A + A + 3 = 5$$
$$2 + 3 = 5$$
$$A + 4 = 5$$

How many ways to make 4?

$$A + A + 2 = 4$$
$$A + 3 = 4$$

How many ways to make 6?

$$A + A + 4 = 6$$
$$2 + 4 = 6$$
$$A + 2 + 3 = 6$$

After the child is comfortable with the combinations, mention to him that if the two Aces are different, say, one is red and the other is black, then show him that there are more than three ways to make 6:

$$A_{red} + A_{black} + 4 = 6$$
$$2 + 4 = 6$$
$$A_{red} + 2 + 3 = 6$$
$$A_{black} + 2 + 3 = 6$$

The child begins to learn the subtle difference between the third and the fourth ways of summing to 6. (Such subtle differences are extremely important in mathematics, in particular, in probabilities — see "Post-Appendix").

By introducing more cards and summing to different totals, a countless variation of this "probability" game can be played. Do not be too concerned over the "correct" answer. Just let the child explore different combinations. ("Probability" as a formal subject in mathematics will not be introduced to the child in school until the higher secondary classes!)

NEVER

force him to continue
"playing" when he wants to
do something else..

Intellectual skills learnt from this game

"Probability" is a difficult branch of mathematics,
and is often neglected or omitted in formal courses
in mathematics in schools and universities. All the
same, it is useful for a child to have a simple
understanding of the different ways in which cards
can be used to arrive at a total (the technical terms
of "combinations" and "permutations" need not
concern us here. Nor is the "correct" answer valuable
at this early stage of a child's introduction to
"Probability").

Simple PROBABILITY

Simple probability is also useful in understanding and playing some card games that may help sharpen math skills e.g. simple gin-rummy (see Appendix). An understanding of probabilities also helps in other games (or helps in the avoidance of some commercial games), so that one is not excessively disadvantaged in games without even realising it e.g. in lotteries and "lucky draws".

It has been said that "the lottery is a heavy tax on those who cannot do math" — meaning that those who do not appreciate the extremely low probability of winning in lotteries and "lucky draws" may end up paying excessive amounts of money to play the lottery. The resulting high gains from operating such lotteries then go to governments (or to private operators who are licensed to operate such games).

Game
16

Simple

LOGIC AND
DEDUCTION

Simple LOGIC AND DEDUCTION

This game gave my two grandchildren aged 4 and 6, endless hours of fun. We call it "Tee Taa Too" (You too can invent your own nonsensical funny sounding names for any of the games, as long as it makes it more fun for the family. Children love funny sounds for all kinds of activities and objects: see e.g. "Lolla-Lee-Lou", "Oompa-Loompas", and "Yonghy-Bonghy-Bo", in the writings by "Dr Suess", Roald Dahl and Edward Lear, respectively).

Start simply. Begin as in the example opposite.

	A	B	C
	(Tee)	(Taa)	(Too)

A	A	B	C
(Tee)	(Tee)	(Taa)	(Too)

			C ?
			(Too)

The child may recognise the pattern
in the cards and put AAAB.
Congratulate him.

At the highest level, mathematics is often regarded as "the science of patterns" rather than "the science of numbers". Many branches of advanced math exhibit beautiful visual and abstract patterns – e.g. topology, fractals, tiling, knots, networks, etc are largely about patterns rather than numbers.

GAME 16

Play the same game a number of times, using different cards for A, B and C. When the child is comfortable, progress upwards:

A	B	C
(Tee)	(Taa)	(Too)

A	A	B	C
(Tee)	(Tee)	(Taa)	(Too)

A			?
(Tee)			

My grandchildren had some difficulties initially when playing with the cards until I started adding the funny sounds. With the aid of the verbal patterns, they saw the visual patterns immediately, and thereafter had no more difficulty even with the more complex ones. This is why it's important for parents/grandparents to be patient, to be creative, to be relaxed and to have fun with their child, and see from time to time such "eureka" moments. Listen:

Tee Taa Too
Tee Tee Taa Too

Isn't it "obvious" that the next line is

Tee Tee Tee Taa Too ?

GAME 16

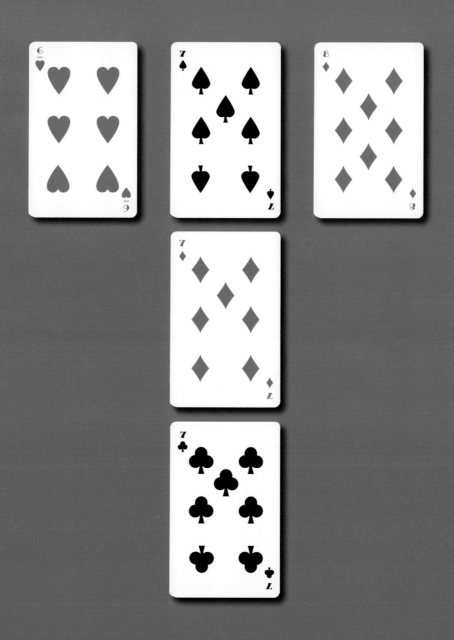

Simple LOGIC AND DEDUCTION

For these games, parents/grandparents must always remember that at least two lines are necessary for the child to deduce what the third line would be. Sometimes, multiple answers are possible for line 3; accept the child's version. (He may have a good explanation for his choice).

As the child enjoys the game, introduce higher intellectual challenges by adding more characteristics into the patterns: e.g.

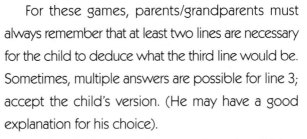

A_{heart}	2_{heart}	3_{heart}	4_{heart}	
A_{spade}	2_{spade}	3_{spade}	4_{spade}	
A_{club}				?
$A_{diamond}$				?

This is just "sequencing" in disguise.

A_{heart}	2_{heart}	3_{heart}	4_{heart}	
2_{spade}	3_{spade}	4_{spade}	5_{spade}	
3_{club}				?
$4_{diamond}$				?

A_{heart}	2_{spade}	3_{club}	$4_{diamond}$
2_{heart}	3_{spade}	4_{club}	$5_{diamond}$
3_{heart}			
4_{heart}			

More complex patterns, when the child is ready.

6$_{heart}$	7$_{spade}$	8$_{diamond}$	
	7$_{diamond}$		?
	7$_{club}$		?
4$_{heart}$	5$_{spade}$	6$_{diamond}$	
		7$_{club}$	?
		8$_{heart}$	?

The use of the fourth suit is a clue for the child that he needs to consider all four suits for filling in the patterns.

More complex patterns, including the use of picture cards, use of odd number cards only, use of even cards only, use of progressively larger number patterns instead of just consecutive cards

(e.g. A	2			
A	2	4		
A	2	4	?	)

Always remember not to tell the child that he is wrong especially as we move to open-ended patterns, as in the last example above. Let the child "do his own thing" and if necessary ask him to explain his choice. You may be surprised by the child's logic

and decision, sometimes in directions that you didn't think about. (Move to simpler patterns if a child does not get the complex ones correct consistently).

After my granddaughter Rebecca, 6, had played "Tee Taa Too" with cards, and enjoyed the many variations that I created for her, she moved upwards to adult IQ books. I found an old book (which I bought as a student at Cambridge some forty years ago) "Check your own IQ". After doing a few of the pictorial/graphical ones with me, she went on to do the rest on her own, in many cases, with little difficulty. Indeed, she used the old book so often that the binding fell apart.

Rebecca's enthusiasm and ability to do most of the pictorial/graphical "tests" in an adult IQ book is testimony to how she learnt Logic and Deduction from "playing cards with Grampa". It makes all the effort and the hours that I spent "playing cards" with her worthwhile.

"Check your own IQ" by H. J. Eysenck, Professor of Psychology at the University of London. He is also the author of numerous other educational psychology books, including "Know your own IQ", "The Scientific Study of Personality" etc.

Simple

LOGIC AND DEDUCTION

Intellectual skills learnt from this game

Notwithstanding perceived shortcomings and criticisms from various quarters leveled at IQ tests, they have and continue to be used extensively for many purposes, including the testing of children for entry in schools, colleges, and universities, for inclusion into special/gifted programs, testing of applicants for jobs, testing of applicants for scholarships, bursaries, grants and other awards. Most of the "logic" tests consists of pattern recognition and arrangements. Simple pattern recognition and deduction with "playing cards" not only give your child immense fun and pleasure, but also a strong foundation, should they encounter such tests later on in school or in life situations. A strong analytical mind coupled with some practice (with a lot of love and fun) will stand your child in good stead in due course.

Game **17**

Simple

ALGEBRA

While J, Q and K can be used to represent 11, 12 and 13 respectively in sequence arrangements, and 10's for "fishing", the "Joker" takes the idea of variability even further. The "Joker" can be used to represent any number from 1 to 10, in "fishing", and any card from 1–K in "SNAP" and "Gin-Rummy".

After the child is comfortable playing "fishing" with multiple cards, adding up to numbers in their 20's, with "picture" cards representing 10's, parents/grandparents can introduce the "Joker" as a "magic card" which can be used for any number that the child likes from 1 to 10. Hence, a "Joker" plus 2 can be 3, 4, . . . all the way to 12 for "fishing". The child will learn this flexibility of the "Joker" very quickly as it helps him win more cards. He will love playing "fishing" with the "Joker".

The "Joker" is also extremely valuable for the different variations of "SNAP", since a "Joker" present will always give "SNAP". The "Joker" adds a lot of fun and laughter to a lot of games. And it teaches the child a lot too. He learns the general concept of a card being able to take on any value, from 1 to 10, and from 1 to 10, J, Q and K in other games (e.g. SNAP or Gin-Rummy).

GAME 17

When the child is comfortable playing with the "Joker" in the different games, he is ready for his first introduction to simple Algebra.

Stick two sheets of "post-it" paper on two cards, and write the signs "+" and "=" on them respectively (see Figure). Then using the "Joker" and a "2", set up the equation (Figure). Ask the child what is the value of the "Joker". If the child is familiar with "fishing" with the Joker, he will have no difficulty saying "8" to the equation opposite. By changing "2" to other numbers, the child learns the variable values for the "Joker".

Progress upwards by changing the number on the right-hand side (RHS) of the equation. Again the child tells you the value of the "Joker". Play this game by changing the numbers on both the left-hand side (LHS) and the right-hand side (RHS). When the child is comfortable, change the "+" sign into "-", and let the child play some more.

Then take another "post-it" paper and write the letter "x" and stick it on the "Joker". Ask him what is the value of "x". Some children can be confused by the "x". Many are not, and can tell you straightaway what the value of "x" is. Congratulate him when he gets the value of "x" correct. He has just done his first exercise in Algebra!

By changing the numbers on both the LHS and the RHS, the child gets to work out the value of "x", practicing his algebra.

Do not rush the child if the child is unsure when you changed from "Joker" to "x". If necessary, remove the "x" and go back to "Joker" until the child is comfortable and consistently correct.

When the child is ready with "x", parents/grandparents may wish to introduce simple multiplication by changing "+" to "×". Point out to the child the difference between "×" for multiplication, and "x" for the romanised alphabet which is used extensively to represent anything unknown.

When the child is comfortable, create another variation by having games with two "x's"

e.g. $2 "x" + 4 = 10$

what is "x"?

Over time, the child learns real algebra which is just a combination of addition, subtraction, multiplication and division involving "x" and multiple "x's".

Progress upwards to 3"x's" when the child is ready and enjoying himself.

Intellectual skills learnt from this game

With this game, the child has learnt the basics of simple algebra which is essentially arithmetic involving the use of "x" ("Joker"), and deducing the value of "x" (Joker). Generally, "x" is used in mathematics for representing an unknown (the mathematical term used is "variable").

The child has informally taken a big step into the wonderful world of algebra, which like arithmetic is a key foundation of mathematics. Indeed arithmetic and algebra are two of the most important pillars of mathematics. And mathematicians use the key principles of arithmetic and algebra routinely in their daily work.

NEVER

scold a child when he doesn't do things right; encourage him.

Game

18

Simple

SIMULTANEOUS EQUATIONS

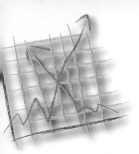

"Playing cards" with our child have brought us to the point when they can do "real math" — simple arithmetic and algebra — the fun way. Most important of all, the child had enjoyed the games, enjoyed playing with parents/grandparents, and enjoyed learning vital, valuable and critical life-long skills. If we have succeeded so far, he would come to like math, or even to love math, as many mathematicians, scientists, engineers and other professionals do.

When the child is comfortable with simple algebra dealing with 2"x's" and 3"x's", create your own variation by introducing another "Joker". (Different packs of cards usually have different "Jokers" or "Jokers" in different colours. If the "Jokers" are the same, improvise with another "post-it" paper and write "y" on top of it.

Go back to the earlier game

$$x + 2 = 4$$

What is x? When the child tells you its 2, congratulate him. Now, ask him to remember that x is 2; then show him:

$$x + y = 5$$

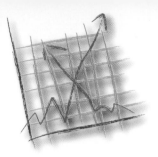

What is y? When he says 3, congratulate him. And don't forget to congratulate yourself also because you have just introduced him to the concept of "2 unknowns" (variables) which is the foundation of "simultaneous equations".

The term "simultaneous equations" is a complex-sounding phrase, and frightens most people including adults who are not comfortable with math. It is an area of math that is normally covered only in secondary school, and often gives children much pain and confusion, leading to their "hating math" in extreme cases.

But for our child, brought up in math through playing cards, "simultaneous equations" is just another game for us to play with him.

Simultaneous equations is just a game played with more than one Joker (usually two) or "x" and "y", if you prefer to use the correct math symbols.

After the child is comfortable with the game above:

$$x + 2 = 4 \qquad x = 2$$
$$x + y = 5 \qquad y = 3\,;$$

use a few more examples by changing the numbers. Over time, the child understands: find the value of x and then find the value of y.

When he is comfortable, try this:

$$y + x = 5$$
$$y - x = 1$$

The child can "guess" that x is 2 and y is 3, recalling from his previous game. Explain to him that if we add the two equations vertically, term by term, we get

$$y + x = 5$$
$$y - x = 1$$

Adding vertically:

$$2y \quad\; = 6$$
$$y \quad\; = 3$$
$$y + x = 5$$
$$3 + x = 5$$
$$x = 2$$

Watch the child closely to see if he has his "eureka" moment — when he understands the concept of adding vertically (similar to his "score-keeping" in "fishing").

If it is not obvious that the concept has "clicked" in the child's understanding, play the game again

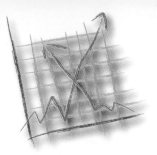

using other "small" digits. Remember not to rush. It is more important for the child to enjoy the game, and get the value of the "Jokers", than to be pressured into doing the sum. While "simultaneous equations" is "math for older children", young children can have fun with it too, if they are ready.

Progress to higher values of x and y only when the child is comfortable, and gets his answer right consistently, showing that he fully understands the basics of "simultaneous equations".

At a later date, when the child is ready, parents/grandparents can create their own variation by using multiple x's and y's. Before long, the child will be comfortable and enjoying his "simultaneous equations" game, including those involving the higher digits.

NEVER

rush the child to "higher" level games; be guided by his preferences.

Intellectual skills learnt from this game

With this last game, the child has traverse an exciting journey through the world of arithmetic, learning the basic skills of pattern recognition and the arithmetic skills of addition, subtraction, multiplication and division. He has also acquire an appreciation of basic concepts in math such as symbols, logic and deduction, sets, probability and simple algebra. More likely than not, he would be known in his class and school as a "math expert". His self-confidence and his self-image would be greatly enhanced, helping in his development, educationally, socially and emotionally.

And all these skills were acquired through "playing cards" and while having fun with parents/grandparents. There are few skills and capabilities that a child cannot achieve when he is properly motivated.

Most important of all, if we can help the child enjoy his attainments and achievements, we help him on his road to the ultimate of education, namely self-motivation to want to learn more, to want to know more, and to want to achieve more.

Have fun, and continue enjoying "playing math" with your child.

Life is good for
only two things:
discovering mathematics,
and teaching mathematics.
Simeon Poisson

All mathematicians share . . .
a sense of amazement
over the infinite depth
and the mysterious beauty
and usefulness of math-
ematics.
Martin Gardner

What science can
there be, more noble,
more excellent,
more useful, more admirably
high and demonstrative
than this of mathematics?
Benjamin Franklin

APPENDIX

Thousands of card games are available in books and on the internet. Not all of them have value for the teaching of math to young child through the play way.

Three of the simplest with direct Play Math value are "SNAP", "Simple Gin-Rummy" and "Are you the King or Are you the Joker?".

"SNAP"

"SNAP" is an extremely simple game that even toddlers can play with great fun and pleasure. Yet with creative variations, it can be played by older children with equal enjoyment (and best of all with immense math benefits).

For the toddler, parents/grandparents can play "matching colours" with the young child. By throwing a card each, "face up" at the same time, the players say "SNAP" when they see two cards of the same colour. Of course, to encourage the child, parents/grandparents can delay saying "SNAP" until the child has said it (you'll be surprised, even very young children loved to "beat" their parents/grandparents). Very soon, if you are observant, you'll notice that the child will "learn" the game, and deliberately delay throwing his card down a fraction of a second after the adult has thrown his. This gives the child a slight advantage of observing the adult's card first before he throws his own and helps him win more often. Allow the younger child this slight advantage as one observes the intricate working of his little mind! For older children, after allowing it for a few rounds, gently encourage the child to throw the card down together. This is an opportunity for encouraging and teaching the value of "fair play".

For older children, create your own variations to add more fun (and math into the game). Variations are countless: they can include:

1. Same suit
2. Sum of the two cards to 10, 11 . . . twenties
3. Difference of two cards by 1, 2, . . .
4. Leaving cards not won on the table, and summing of one of the two "fresh" cards together with one of the "table" cards to 10, 11, . . . twenties.
5. As in 4, but with a difference of 1, 2, or 3.
6. Multiplication: taking 8, 12, 16, 18, etc. (two 4 or four 2's; three 4 or four 3's, etc., saying the multiplication out aloud).
7. Division: e.g. taking two cards, 8 and 2 with the player saying "8 divide by 2 is 4", etc.
8. "Greater than" — saying "SNAP" and pointing at the person with the "higher" card or saying "8 is greater than 4" etc.
9. "Less than" — saying "SNAP" and pointing to the person with the "lower" card, etc.
10. "Open-ended multiplication SNAP" — looking at the two cards and saying "SNAP 6 × 3 = 18" etc.

"SIMPLE GIN-RUMMY"

Gin-Rummy is an adult game that can be played by children with cards. Its basic principle is similar to that of "mahjong", "four-colours" as well as many other card games.

Simple gin-rummy (our grandchildren love to call the game mini-gin) can be used to help a child exercise his memory whilst learning about sets, sequencing, probability, etc.

The game can also be played with multiple players using two or more packs of cards if necessary.

For two players (e.g. child and Grampa), each player is given seven cards, with one additional card face-up on the table. The child learns to arrange the cards in "tentative sets", putting the cards of the suit together in numerical sequence (e.g. 6(H), 7(H) and 9(H) ignoring the absence of a complete running sequence of three or more cards) or putting the cards of the same quantities together (9(H) 9(S) 9(D)). The child also learns to manipulate the seven cards in his hand, "fan" it out and "read" the integers in the corner (you'll be surprise how this physical act, which is so simple to adults, is extremely difficult for young children). If another adult is present, he can lend the child a hand and help him in the initial stages. If no other adult is available, begin by encouraging the child to put his cards face-up on the table by doing the same yourself. Organise both his cards and your own in "tentative sets" and tell him what cards he would like to take up (if they are available) to complete his "set of three".

The adult can begin the game by taking the "table card" if it is useful for you to help in your "set". If not, ignore the "table card" and draw a new card from the "reserve stack". If the card is of no value to you, throw it down and this becomes the "current table card".

Help the child to see if the "current table card" is of use to him. If so, explain to him how the card can help him to build his "tentative set"; then throw out a card that is "least valuable". Explain to the child why it is "least valuable". If the "current table card" is of no use to the child, encourage him to draw a new card from the "reserve stack" and follow the same process as before.

In the initial games, do not overload the child with tactics and strategies. Just help the child to "play" the game. The player who gets "two sets of three cards" and a set of "doubles" wins.

Although the written instruction sounds complicated, most children learn extremely quickly and often after two or three games (with a lot of noise

128

and excitement and fun) they will know how to "play".

When the child is comfortable, move upwards to 10 cards each, with three "sets of three" and a "double" to win. Our grandchildren call this game "normal-gin". After the child is familiar with the game, explain the different probabilities e.g. 2 and 3 "waiting" for A and 4 is better than A and 3 waiting for 2 only (two chances to complete the set in the former, as compared to one in the latter). This is about probability, and soon the child understands that "waiting" from two "end cards" are better than waiting for a "middle card".

When the child can play reasonably well, encourage the child to keep track of the cards that have already been discarded on the table. He should avoid "waiting" for cards that had been discarded earlier in the game as the probability of such cards coming up later is diminished. (You'll be surprised (pleasantly!) when his memory proves to be better than grandparents'.)

If you are a good mahjong player, you can help the child learn more tactics and strategies; but these are not necessary for the child's enjoyment or learning and understanding some of the fundamentals of mathematics.

Child's Choice

Remaining Cards

First
Possibility

Second
Possibility

Third
Possibility

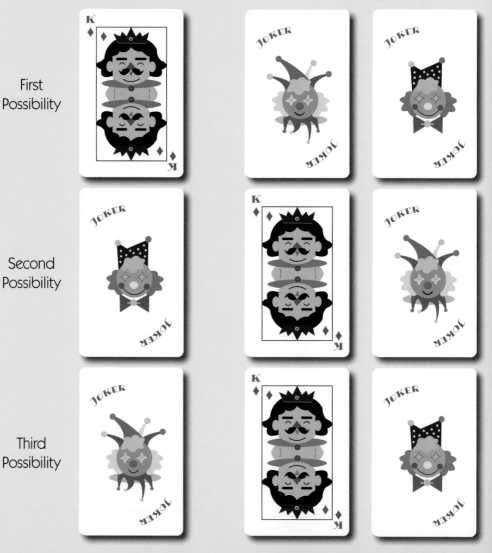

OR ARE YOU THE **JOKER?**

This is a super-simple game that children, including the very young, love to play. It can be hilarious, and gives children (and parents/grandparents) a lot of fun. I created this game by modifying it from a well-known TV game that was popular in many countries in the 60's.

1. Take a King (or a Queen if you are playing with a girl and wish to be politically correct) and two Jokers. (It is assume for this game that the Joker is less preferred than the King).

2. Shuffle the three cards and place them face-down.

3. Ask the child to select one of the cards, and keep it face-down.

4. The adult picks up the two remaining cards, looks at them, and discards the Joker — face-up.

5. Then put back your remaining card, face-down.

6. Ask the child if he wishes to swap his choice card with yours.

7. Now say together — "Are you the King or Are you the Joker?"

8. Ask the child to open his card and you open yours.

   (The child's can say "I am the King and you are the Joker" if his card is a King.)

Questions:

Should the child keep his chosen card?
Or should he swap? Always? Sometimes?
Does it make a difference?
What is the correct strategy?
What is the mathematical principle behind the decision?

POST-APPENDIX

This book was written to help parents/grandparents teach math to their pre-school and primary school children the fun way.

Nevertheless, I am sure some parents/grandparents are sufficiently stimulated intellectually by the questions in "Are you the King or Are you the Joker?", and would like to know the answer, and how it was arrived at.

I have therefore included this unusual "Post-Appendix", which is not part of the book proper (hence the term Post-Appendix). Parents/grandparents who do not wish to, do not have to read any further. If you wish to read on, it's for your own entertainment and edification.

ARE YOU THE KING

As I indicated earlier, probability is one of the more difficult branches of mathematics, in school, in university, as well as in real life applications. While there are obviously general principles to apply, minor variations in the conditions of the problem can give widely different results (answers). Hence very careful analysis of a problem is necessary before the problem can be properly structured in "probability terms".

Correct Answer for the game "Are you the King or Are you the Joker?"

The child should always swap his chosen card with the adult's card (after the adult had thrown away the Joker).

This way, he will double his chance of being King.

This super simple game, which I have created for my grandchildren, can be played with great enjoyment by even the very young, but maybe without full understanding of its implications.

The game is a variation of a common game first recorded in the 19th Century, and appears in many different forms in statistics textbooks.

OR ARE YOU THE JOKER?

Professor Ian Stewart in his book "The Magical Maze" gave an entertaining account entitled "Marilyn and the Goats", of how the game was played on TV, and how an embarrassing and controversial incident ensued thereafter. Briefly, there was once a popular TV game where a contestant on stage was offered three doors, with prizes behind them. Behind one door was a car. Behind each of the other two doors was a "goat", the booby prizes. The contestant must select one, and only one door. Thereupon, the TV presenter would open a door (he knows what's behind each of the doors), and throws away the "goat". Then, he asked the contestant if he wishes to change his choice, to swap his door for the presenter's remaining door.

Marilyn vos Savant is reputed to be the holder of the highest recorded IQ in the "Guinness Book of World Records". In 1990, in her newspaper column, her answer to the above riddle was that the contestant should always swap with the presenter because after swapping, his probability for winning the car would be doubled.

Marilyn's answer was greeted by widespread criticism and a huge deluge of angry letters, including many from professional mathematicians and statisticians, claiming that she was wrong. The furore was so intense that the controversy made it to the front page of the august "The New York Times". A further explanation in a subsequent column convinced some readers but not all. Eventually, after further detailed explanation, and real runs of the problem by thousands of school children, the controversy died down, with Marilyn proven to be correct after all.

Despite her high IQ, Marilyn was unable in her earlier columns, to explain to the satisfaction (and understanding) of her readership how the problem should be addressed. This reflects the difficulty of "probability" problems.

I shall attempt to explain the King/Joker riddle simply, in the context of the games in Section I, which we encourage parents/grandparents to play with their children.

In Games 2, 3, and 4, one of the key concepts that we wanted our children to acquire was the ability to FOCUS on the relevant characteristic, and IGNORE all the other distracting features which can add to the confusion. In Game 2, the child focused on colour, and ignore the quantities, suits, etc. In Game 3, he focused on the shapes (suits) and ignore the colour, numbers etc.

In the Game "Are you the King", the answer can be easily understood if we focus on the probability (or chance) at the time of selection.

Child's Choice	Remaining Cards
One Card	Two Cards
"1 in 3"	"2 in 3"
probability of King	probability of King

(see Figure on p. 130). When the child selects one card at random, the probability of having selected the King is "1 in 3". The probability of the remaining cards having the King is "2 in 3" (because there are two cards!). If the child is asked if he would like to swap his one card for both of the adult's two cards, the answer would be obvious to all — YES! Swap, and double his chance of being King. That is the nature of the problem in "probability" terms.

The subsequent actions, are a distraction to the problem, because they do not change the probabilities of "1 in 3" for the child's one card and

"2 in 3" for the adult's two cards. Nevertheless, an explanation for why subsequent actions are a distraction in "probability" terms is useful.

Because the adult has to two cards and therefore a "2 in 3" chance of being King, he will always have at least one Joker also. By looking at his cards and throwing away a "Joker" he has not changed the child's probability of "1 in 3" or his own probability of "2 in 3".

Child's Choice	Adult's Card
One Card	Remaining card + Joker (opened and thrown away)
Still "1 in 3" probability of King	Still "2 in 3" probability of King

After a Joker is thown away, the adult's remaining card now has a "2 in 3" probability of being King, compared to the "1 in 3" probability for the child's card.

In diagrammatic form (Figure on p. 130):

Possibilities for Child's One Card	Possibilities for Adult's Two Cards	
1. King	Joker(A) + Joker(B)	
2. Joker(A)	King + Joker(B)	
3. Joker(B)	King + Joker(A)	

After the adult had looked at his two cards, and had thrown away a Joker, the possibilities are as follows:

1. King   Joker (either A or B, as it does not matter)

2. Joker(A)   King

3. Joker(B)   King

The child has a "1 in 3" chance of being King. The adult has a "2 in 3" chance of being King. Therefore the child should always swap with the adult, and in the process, double his chance of being King.

Are you convinced now?

New Game

If you understand fully the problem in terms of probability analysis and are fully convinced that the child should always swap, let me give a slight twist to the problem to illustrate the point about minor changes in conditions giving significantly different result's in probability problems.

If after the child had chosen his card, the adult <u>without</u> looking at his cards, now randomly throws away one of his cards,

What should the child do now?

1. Always swap?

2. Never swap?

3. No difference?

If you understand your "probability" analysis, then you'll know that the third answer is correct — No difference. Both the child's card and the adult's remaining card have the same "1 in 3" chance of being King. The card that was randomly thrown away also has a "1 in 3" chance of being King.

Convinced ??

Another New Game

Take three identical Kings of Hearts and three identical Jokers from three packs of cards. Using double-sided sticky tape (or blu-tack), stick two Kings back to back so that you have a double card with King on both sides. Do the same for the two Jokers. Then stick the remaining King with the remaining Joker so that the third double card has King on one side and Joker on the other.

To play the game, put the three cards in a non-transparent bag. Ask the child to put his hand into the back, mix the cards, and draw out one card <u>without</u> looking at it. Place the card on the floor (or table). Then ask the child — "Is it a King or Is it a Joker on the other side of the card?"

What about you? Do you know the answer? What is your guess? How do you win at this game?

The overwhelming majority of players — both adults as well as children — will think that there is a 50/50 chance that the hidden picture (facing the floor) is either King or Joker. So they conclude that it makes no difference what the guess is.

The correct answer — there is a 2-in-3 chance that the picture on the other side is the same as the picture facing up i.e. if there is a King facing up, then there is a 2-in-3 chance that there is also a King facing down.

Can you figure out why?

If you are not convinced, play the game thirty times; then the expected occurrence of same picture on both side is twenty times, and that of the non-identical pictures is 10 times.

Are you convinced now?

Do you know why now?

BIRTHDAY
COINCIDENCES

Another example of difficulties of probability problems is given by the classic group of "birthday coincidences" problems.

Two persons having the same birthdays

Imagine you are at a party. How many people should there be before there is a "better-than-50%" chance of two people having the same birthday (assume there are 365 days in a year)?

If you test the question on your friends (unless they have read the answer beforehand), you are likely to get all kinds of guesses, with the majority (especially the more thoughtful and math-oriented ones) clustering around 183 (half of 365 days). Almost certainly, the overwhelming majority will be wrong. The correct answer is 23 people.

Analysis

Let us assume person A arrived at the party first. Then came person B. It is extremely unlikely that B will have the same birthday as A, as there are 364 days for B

143

to have his birthday on, without having it coincide with A's birthday. The probability of B <u>not</u> having the same birthday as A is 364/365. Person C turned up next. There are 363 days for C to have a birthday without coinciding with either A or B. The probability for C not having the same birthday as A and B is

$$\frac{364}{365} \times \frac{363}{365}$$

For persons B, C, D etc <u>not</u> having the same birthday as the other earlier arrivals, the probabilities are

$$B = \frac{364}{365} \qquad\qquad = 0.997$$

$$C = \frac{364}{365} \times \frac{363}{365} \qquad\qquad = 0.992$$

$$D = \frac{364}{365} \times \frac{363}{365} \times \frac{362}{365} \qquad = 0.984$$

....

....

$$\begin{array}{ll} \text{Person W} \\ \text{(23}^{\text{rd}}\text{ arrival)} \end{array} = \frac{364}{365} \times \frac{363}{365} \times \cdots \frac{342}{365} = 0.493.$$

With Person W, the 23rd person to arrive, the probability of him not having the same birthday as the preceding 22 people is 0.493. In other words, with 23 people, the "converse probability" namely the probability that any two people at the party having the same birthday is 0.507 or 50.7%.

As easy way to remember the answer for this problem is to think of a football match with 11 players on each side and a referee. Each time a match is played with all 23 participants present, there is a "better-than-50% chance that two of them will share a birthday.

Are you amazed at how small the number is? Are you convinced?

One other person having the same birthday as you.

Imagine you are at a party again. How many other people should there be before there is a "better-than-50%" chance that one of them will have the same birthday as you.

If you test the question on your friends (again, unless they have read the answer beforehand) you will get all kinds of guesses again. Again, most will cluster around 183 (half of 365) if they have not heard or known the answer to the earlier problem. If this second problem is posed after the answer of 23 had been given to the earlier one, guesses will now cluster around numbers slightly more than 23 (e.g. 24, 25 etc).

The correct answer is 253.

Analysis

Let us assume you are the host A, the first person at the party. Then came person B. The probability of B not having the same birthday as you is 364/365. Then came person C. The probability of C not having the same birthday as you is also 364/365, since there are still 364 days for him to have his birthday without coinciding with yours.

Hence the probability of both B and C (two people not having the same birthday as you) is 364/365 × 364/365.

The probabilities for the number of people not having the same birthday as you are:

For 1 person : $\dfrac{364}{365}$ = 0.997

For 2 people: $\left(\dfrac{364}{365}\right)^2$ = 0.994

For 3 people: $\left(\dfrac{364}{365}\right)^3$ = 0.992

For n people: $\left(\dfrac{364}{365}\right)^n$

For 253 people: $\left(\dfrac{364}{365}\right)^{253}$ = 0.4995

When 253 guests had arrived at the party the probability of all of them not having the same birthday as you is 0.4995. The "converse probability" i.e. one of them having the same birthday as you is 0.5005 or 50.05%.

This is a very different answer to the earlier problem because the condition that the other guests must have the same birthday as you (one single specific individual) is a very severe constraint. Even if two other guests have the same birthday, it does not satisfy the original condition of having the same birthday as you.

BIRTHDAY COINCIDENCES

The two "birthday coincidence" problems illustrate once again the nature of probability problems — namely minor differences in conditions can make for significantly large differences in results.

An easy way to remember the answer for this problem (253 other people) is to think of a party host (you) having a party with guests from 11 football matches, each of them with their full complement of 23 participants, (11 × 23 = 253).

Notwithstanding its "difficulties", Probability is an extremely important branch of mathematics and is vitally critical in the commercial world, including the gaming, insurance, investment and financial industries.

NEVER

Some Reminders

NEVER force a child to "play cards" when the child prefers to play other games e.g. with animals, boxes, dolls, lego, mechanical sets, etc. ⊗

NEVER discourage a child. ⊗

NEVER tell a child he is wrong; show him the right way. ⊗

NEVER scold a child when he doesn't do things right; encourage him. ⊗

NEVER rush the child to "higher" level games; be guided by his preferences. ⊗

> Do not worry too much about your difficulties in mathematics,
> I can assure you that mine are greater still.
> **Albert Einstein**

ALWAYS

ALWAYS congratulate the child when he has completed his "game" and "made progress". ✔

ALWAYS encourage the child. ✔

ALWAYS be observant to see where the child may have misunderstood and is unable to "play", and help him. ✔

ALWAYS ensure that the child is enjoying himself. ✔

ALWAYS be attentive to the child's preferences; they give indications of the child's level of learning and mastery of the underlying math that you are attempting to teach in the play way. ✔

ALWAYS invent strange funny nonsensical sounds or names for your games. (It makes the time spent playing cards with parents/grandparents that much more fun and enjoyable. And you'll be greatly rewarded by the progress made in math by your child.) ✔